Zahid Khokhar

Genteel Evasion However

Zahid Khokhar

Genteel Evasion However

Crossbreeze

Blessed Hope Publishing

Imprint

Cover image: www.ingimage.com

Publisher:
Blessed Hope Publishing
is a trademark of
International Book Market Service Ltd., member of OmniScriptum Publishing Group
17 Meldrum Street, Beau Bassin 71504, Mauritius
Printed at: see last page
ISBN: 978-613-7-94700-5

Title: Genteel Evasion However

Subtitle: Crossbreeze

Author: Zahid Hafeez Imam Khokhar

Table of Contents

Preface

The concise work can be found here in this hurriedly written manuscript. First of all a set of introductions covering the topics of human generation existence and their reason of existence on the earth are highlighted. Each of these has many articles combined in separate volumes. General paragraphs are described in the introduction and the mentioned articles and book there in those paragraphs are works. These are not merely a work of sitting but a combination of life experiences, vast reading, travelling around a lot and thinking with knowledge spread within varieties of books of history and modern sciences. Science is a knowledge of the unknown that is always kept in the mind. What the history tells and what are the thoughts of present minds are both valuable.

Humans are developing themselves more and more along the way keeping themselves and the environment more comfortable for their living. Machines, chemical plants and work areas are needed to be developed presenting more ease to human life. That is the wisdom and knowledge of humans which God has revealed to them time by time with their struggles which are achieving success in this way. This mastery humans are getting from the life and death created by God and along with this humans are to take very careful care and consideration about the messages of God which have reached them by the chain of Prophets of God. The essence of previous messages is contained in the present Book of God of Prophet of Islam so that humans and jinns can succeed in the worldly life examination of God to be in Paradise with having more good counts than evil counts.

Introduction

Certs compulsion wherever any human is the life and the death and the life after death. The human has to live on the earth for a specific time and then the death towards God, human has to face math of good and evil for judgment to hell or not. Moderate distinct whatever should always be aimed.

The concern of every human is a comfortable life here on earth because the humans actually belong to Paradise. It is in their genes that they are looking for the taste of Paradise and every one struggles for that more or less. That is what God has put in them : evil and good and let them live on the earth clearing the path of evil as well as path of good. Reminders from God are presented before humans by the Prophets of God that where the path of evil leads and where the path of good leads. Backing origin whenever should always be remembered.

Hurry-make humans are for forever and at the day of judgment part day of God, firstly bold humans (and the jinns) on truth will be in paradise and then separations will be lined up. Moreover some would have to wait on for sometime.

God remembers, forgets, mentions, seeks and waits for but never takes sleep or catches catnap as God likes, his own styles to remind. Who wants satisfaction finds it. God taught creations and those are on. The earth had been created for humans to live upon to then come back to before God. Some of the contents of the book are hard, some are soft, some are rolling, some are watering, some are cautioning and some are poetry. In razzle or in cell, consequences might differ. Ratiocination is inclusive. Death and life are running after each other and at last life would have to survive. That's what the humans are for.

All knowledge is from the treasures of God and treasures of God are limitless. Creations of God whether seen or not seen due to far distance or due to small size exist and humans reach is expanding towards both sides. The humans at their soul-state life confirmed God before God being their God. Here on the earth they are to explore earth life recognizing God. The versatile mentioned articles are contained in the book for the reader to get the knowledge of doing research and explanations, keeping in mind to moderate rather than ignite the faced realities.

God has created life and death to testify humans and jinns for doing the best good. Life on earth is in the propagation stage. The initiation and termination on life on the earth are mentioned herein. God is asking both humans and jinns to count blessings and boons upon them denied - not found yet - so to remember God. Good and evil struggle is on and these will be separated on the day of judgment. The articles contained are related to what has happened and what is happening and how the humans (and the jinns) are to get towards what God is asking - here on the high.

'Present nature before wisdom. Left it did, you do'. The lines and paragraphs contained in this book describe somewhat separation feelings. Humans are on the earth to explore and use earth during their lifetime. What is known and what is not known are encountered here on the earth for humans. Reading nature is a task for humans to reach back to from where they belong. The teachings of Prophets of God are to remember God who examines each and everyone in his own style. Humans are equipped with knowledge and codes of nature explored to get benefited from nature provided sources. Good and evil both exist on earth and would be separated on the day of judgment.

Humans were created to live on the earth for a fixed period of time to explore and conquer the earth-system. They were put in paradise for a temporary stay to get knowledge of things and how to live where they are to come back by doing good on the earth-life avoiding slip for hell-punishment. They all at their soul-state life confirmed God being their God while at the sky-system. The short articles contained in this book talked about the start of human creation by God when God was in a hurry to create humans authoritative on the earth where there are many creations on and around the earth, and the attachments and explanations to that pointing past, present and future.

Mission

No one has any permission before God to 'shefarush' anyone except whom God has permitted. 'Shefarush' is an act of creation for other creation and its acceptance is in the hand of God. 'Shafaat' is a permission of 'shefarush' granted by God with an already firm acceptance of God. Many creations of God will 'shefarush' at Qiyamat while everyone would be under accountability upon its actions during worldly lifetime. Beliefs of God, Angels of God, Books of God, Prophets of God and Day of God, and the lifetime actions of the humans and the jinns from the day first to last day at worldly life which are continuously under observation of each by Angels of God for God shall be presented to each of whom those were. Parts of the last and final Book of God will be guardians upon their reciters. At Qiyamat each will be behind whom each had worshiped here in the world. Followers of Sun will follow Sun, followers of Moon will follow Moon and followers of any other will follow that. God will be recognised in a second attempt by followers of God who will be in a known shape then.

Quadrilateral

No one has any permission before God to 'shefarush' anyone except whom God has permitted. 'Shefarush' is an act of creation for other creation and its acceptance is in the hand of God. 'Shafaat' is a permission of 'shefarush' granted by God with an already firm acceptance of God. Many creations of God will 'shefarush' at Qiyamat while everyone would be under accountability upon its actions during worldly lifetime. Beliefs of God, Angels of God, Books of God, Prophets of God and Day of God, and the lifetime actions of the humans and the jinns from the day first to last day at worldly life which are continuously under observation of each by Angels of God for God shall be presented to each of whom those were. Parts of the last and final Book of God will be guardians upon their reciters.

At Qiyamat each will be behind whom each had worshiped here in the world. Followers of Sun will follow Sun, followers of Moon will follow Moon and followers of any other will follow that. God will be recognized in a second attempt by followers of God who will be in a known shape then.

Dome

No one has any permission before God to ‘shefarush’ anyone except whom God has permitted. ‘Shefarush’ is an act of creation for other creation and its acceptance is in the hand of God. ‘Shafaat’ is a permission of ‘shefarush’ granted by God with an already firm acceptance of God. Many creations of God will ‘shefarush’ at Qiyamat while everyone would be under accountability upon its actions during worldly lifetime. Beliefs of God, Angels of God, Books of God, Prophets of God and Day of God, and the lifetime actions of the humans and the jinns from the day first to last day at worldly life which are continuously under observation of each by Angels of God for God shall be presented to each of whom those were. Parts of the last and final Book of God will be guardians upon their reciters. At Qiyamat each will be behind whom each had worshiped here in the world. Followers of Sun will follow Sun, followers of Moon will follow Moon and followers of any other will follow that. God will be recognized in a second attempt by followers of God who will be in a known shape then.

King Muhammad Jahangir advised to make his tomb without a dome.

Fourteen

No one has any permission before God to 'shefarush' anyone except whom God has permitted. 'Shefarush' is an act of creation for other creation and its acceptance is in the hand of God. 'Shafaat' is a permission of 'shefarush' granted by God with an already firm acceptance of God. Many creations of God will 'shefarush' at Qiyamat while everyone would be under accountability upon its actions during worldly lifetime.

Beliefs of God, Angels of God, Books of God, Prophets of God and Day of God, and the lifetime actions of the humans and the jinns from the day first to last day at worldly life which are continuously under observation of each by Angels of God for God shall be presented to each of whom those were. Parts of the last and final Book of God will be guardians upon their reciters. At Qiyamat each will be behind whom each had worshiped here in the world. Followers of Sun will follow Sun, followers of Moon will follow Moon and followers of any other will follow that. God will be recognized in a second attempt by followers of God who will be in a known shape then.

The Prophet of God Muhammad RasoolaAllah- peace be upon him and salaam - pointed to the Moon on the fourteenth night saying that to the companions present there with him you will be seeing your Rub like this on the day of Qiyamat.

Quar

No one has any permission before God to 'shefarush' anyone except whom God has permitted. 'Shefarush' is an act of creation for other creation and its acceptance is in the hand of God. 'Shafaat' is a permission of 'shefarush' granted by God with an already firm acceptance of God. Many creations of God will 'shefarush' at Qiyamat while everyone would be under accountability upon its actions during worldly lifetime. Beliefs of God, Angels of God, Books of God, Prophets of God and Day of God, and the lifetime actions of the humans and the jinns from the day first to last day at worldly life which are continuously under observation of each by Angels of God for God shall be presented to each of whom those were.

Parts of the last and final Book of God will be guardians upon their reciters. At Qiyamat each will be behind whom each had worshiped here in the world. Followers of Sun will follow Sun, followers of Moon will follow Moon and followers of any other will follow that. God will be recognized in a second attempt by followers of God who will be in a known shape then.

The Prophet of God Muhammad RasoolaAllah- peace be upon him and salaam - pointed to the Moon on the fourteenth night saying that to the companions present there with him you will be seeing your Rub like this on the day of Qiyamat.

The Prophet of God Muhammad RasoolaAllah - peace be upon him and salaam - and his companions once was in a journey and slept a prayer whereupon the Prophet of God Muhammad RasoolaAllah - peace be upon him and salaam - told them that they were slept by the act of Satan. (Satan has been given permissions by God to do so.)

Fourth Khalifa with his wife slept a prayer at night and upon asking told that God had made them sleep. Upon that the Prophet of God Muhammad RasoolaAllah - peace be upon him and salaam - tapped his thigh and said ‘the human is quarlsome’ - the verse was revealed before or then-. And God knows the best.

Grant

No one has any permission before God to 'shefarush' anyone except whom God has permitted. 'Shefarush' is an act of creation for other creation and its acceptance is in the hand of God. 'Shafaat' is a permission of 'shefarush' granted by God with an already firm acceptance of God. Many creations of God will 'shefarush' at Qiyamat while everyone would be under accountability upon its actions during worldly lifetime. Beliefs of God, angels of God, Books of God, Prophets of God and Day of God, and the lifetime actions of the humans and the jinns from the day first to last day at worldly life which are continuously under observation of each by Angels of God for God shall be presented to each of whom those were.

Parts of the last and final Book of God will be guardians upon their reciters. At Qiyamat each will be behind whom each had worshiped here in the world. Followers of Sun will follow Sun, followers of Moon will follow Moon and followers of any other will follow that. God will be recognized in a second attempt by followers of God who will be in a known shape then.

The Prophet of God Muhammad RasoolaAllah- peace be upon him and salaam - pointed to the Moon on the fourteenth night saying that to the companions present there with him you will be seeing your Rub like this on the day of Qiyamat.

The Prophet of God Muhammad RasoolaAllah - peace be upon him and salaam - and his companions once was in a journey and slept a prayer whereupon the Prophet of God Muhammad RasoolaAllah - peace be upon him and salaam - told them that they were slept by the act of Satan. (Satan has been given permissions by God to do so.)

Fourth Khalifa with his wife slept a prayer at night and upon asking told that God had made them sleep. Upon that the Prophet of God Muhammad RasoolaAllah - peace be upon him and salaam - tapped his thigh and said 'the human is quarrelsome' - the verse was revealed before or then-. And God knows the best.

Sleep is like a little death.

Qualified

God reminds the humans that they all are always in vain and loss unless otherwise they become believers, act good deeds and spread truth and patience. The first pair of humans was also taught a prayer of avoiding loss to recite. God revealed his words, sentences and books to his Prophets towards their respective peoples and nations. God selected arabic language for his last and final book to be revealed down on his last and final Prophet in the land of the city Makkah of people having arabic language, God told that the book is revealed in arabic so that the people get wisdom - from the book. Arabic language was by the time passed had become qualified to be chosen. The people were proud of their language development upon other worlds' languages. God suppressed their wording and writings by revealing His Book to His Prophet. Arabic language was by the time passed had become so qualified to be chosen. From the time of the Prophet till including the day of appearing before God, the peoples are invited to get knowledge and wisdom from the Book of God revealed to the Prophet of God to become righteous. The Prophet of God had explained and warned people to do right and good deeds otherwise punishment might surrender them here in this world or in the hereafter, there before God. Arabic will be the Paradise language.

Qual

God reminds the humans that they all are always in vain and loss unless otherwise they become believers, act good deeds and spread truth and patience.

The first pair of humans was also taught a prayer of avoiding loss to recite.

God revealed his words, sentences and books to his Prophets towards their respective peoples and nations.

God selected arabic language for his last and final book to be revealed down on his last and final Prophet Muhammad RasoolaAllah - peace be upon him and salaam - in the mubarak land of the city Makkah of people having Arabic language, God told that the book is revealed in Arabic so that the people get wisdom - from the book. Arabic language was by the time passed had become qualified to be chosen.

The people were proud of their language development upon other worlds' languages. God suppressed their wording and writings by revealing His Book to His Prophet Muhammad RasoolaAllah - peace be upon him and salaam -.

Arabic language was by the time passed had become so qualified to be chosen. From the time of the Prophet till including the day of appearing before God, the peoples are invited to get knowledge and wisdom from the Book of God revealed to the Prophet of God to become righteous.

The Prophet of God Muhammad RasoolaAllah - peace be upon him and salaam - had explained and warned people to do right and good deeds otherwise punishment might surrender them here in this world or in the hereafter, there before God.

Arabic will be the Paradise language.

Mubarak

Reader per writer key haalat, jab us nay likha ho, aur writer ke shakhseyat ka asar parta hay. Aisa kaha jaata hai. Aik writer apnee halat yay likhaain ~ kay jab unhoon nay yay likha us waqt doctoroon hakeemoon nay mind ka kaam karnay say mana kiya. Mubarak waqt.

Apni shakhseyat kay baray main yay likhtain hain ~ kay elders kehtay rahay likho likho aur wo sharminda hotay rahay aur likhna delay hota raha.

Reader

What is written and what the personality of the writer is both affect. This is spread within a group to hinder people to read other views and to get knowledge rise. Then what happens to these people they become strict and stick to that only writing they are fed and other's everything they thought is not worthy. There-written in a book that when the doctors stopped to do mind work then the presented work was compiled thinking of an opportunity of mubarak time. The elder(s) had asked again and again for the year around yearly but the work was delayed so long so.

Celebration

'Aqeeqah' is an act of sacrifice on the birth of a child. It is celebrated on the next week, the repeated same day, that is the eighth day of birth.

'Salaam on the day he born'

It may be repeated weekly on the day of birth or monthly or annually but once in a life is a very highly demandable act from believers.

'Pray for God and sacrifice'

Calling upon a meal on the day is not necessary, however, it may vary from person to person and family to family upon their work status in the society and relationships.

Aqeeqah

'Aqeeqah' is an act of sacrifice on the birth of a child. It is celebrated on the next week, the repeated same day, that is the eighth day of birth.

'Salaam on the day he born'

It may be repeated weekly on the day of birth or monthly or annually but once in a life is a very highly demandable act from believers.

'Pray for God and sacrifice'

Calling upon a meal on the day is not necessary, however, it may vary from person to person and family to family upon their work status in the society and relationships.

Salaam on the day of birth, on the day of death and on the day of getting up again

'Pray for your Rub and Nahar'

Usual

Usually Muhammad Rasool of God 'Muhammad RasoolaAllah - peace be upon him and salaam -' was regular to fast on mondays and thursdays and told people about alternative day fast of Daoud Nabi of God. Muhammad Rasool of God also told people to fast on thirteen, fourteen, and fifteenth of Moon each month and that would be like fasting whole month. Muhammad Rasool of God told that the monday was his birth day and the fast was in a thanks reply to God.

Sixty three per cent

Muhammad Rasool of God 'Muhammad RasoolaAllah - peace be upon him and salaam -' sacrificed sixty three times on his pilgrimage whereas hundred was completed by fourth Khalifa, his son in law. The worldly earth age of Muhammad Rasool of God from birth to death towards God is sixty three calendar years. One sacrifice is compulsory at pilgrimage, for one who has intentions. An eid day prayer was prayed the next day to the declared eid day by breaking fast due to information delay. Prayer of wealth -zakat- for this year and coming year was collected in advance.

Prayers

'Aqeeqah' is an act of sacrifice on the birth of a child. It is celebrated on the next week, the repeated same day, that is the eighth day of birth.

'Salaam on the day he born'

It may be repeated weekly on the day of birth or monthly or annually but once in a life is a very highly demandable act from believers.

‘Pray for God and sacrifice’

Calling upon a meal on the day is not necessary, however, it may vary from person to person and family to family upon their work status in the society and relationships.

Salaam on the day of birth, on the day of death and on the day of getting up again

'Pray for your Rub and Nahar'

Usually Muhammad Rasool of God was regular to fast on Mondays and Thursdays and told people about alternative day fast of Daoud Nabi of God. Muhammad Rasool of God also told people to fast on thirteen, fourteen, and fifteenth of Moon each month and that would be like fasting whole month. Muhammad Rasool of God told that the Monday was his birth day and the fast was in a thanks reply to God.

Muhammad Rasool of God sacrificed sixty three times on his pilgrimage whereas hundred was completed by fourth Khalifa, his son in law. The worldly earth age of Muhammad Rasool of God from birth to death towards God is sixty three calendar years. One sacrifice is compulsory at pilgrimage, for one who has intentions. An Eid day prayer was prayed the next day to the declared Eid day by breaking fast due to information delay. Prayer of wealth -zakat- for this year and coming year was collected in advance.

Salaams and Prays

God be mercy on him

God be pleased with him

Salaam on him

May God be mercy on him

May God be pleased with him

May Salaam on him

'Salaam on the day he born'

'Salaam by Merciful God'

God be mercy on you

God be pleased with you

Salaam on you

God be mercy on me

God be pleased with me

Salaam on me

God be mercy on us

God be pleased with us

Salaam on us

God be mercy on ...

God be pleased with ...

Salaam on ...

The prayers are set for different ranks so to distinguish the personality. However, these set achievements are required for each and every person of ummah.

[Mercy, Please, Salaam]

However!

God be mercy on him

God be pleased with him

Salaam on him

May God be mercy on him

May God be pleased with him

May Salaam on him

'Salaam on the day he born'

'Salaam by Merciful God'

God be mercy on you

God be pleased with you

Salaam on you

God be mercy on me

God be pleased with me

Salaam on me

God be mercy on us

God be pleased with us

Salaam on us

God be mercy on ...

God be pleased with ...

Salaam on ...

The above mentioned prayers are set for different ranks so to distinguish the personality. However, these set achievements are required for each and every person of Ummah.

Mercy, Please, Salaam

God be mercy on him

God be pleased with him

Salaam on him

May God be mercy on him

May God be pleased with him

May Salaam on him

'Salaam on the day he born'

'Salaam by Merciful God'

God be mercy on you

God be pleased with you

Salaam on you

God be mercy on me

God be pleased with me

Salaam on me

God be mercy on us

God be pleased with us

Salaam on us

God be mercy on ...

God be pleased with ...

Salaam on ...

The above mentioned prayers are set for different ranks so to distinguish the personality. However, these set achievements are required for each and every person of Ummah.

Turn

Turning the face aside

When something faced disliked

When eyes can not bear or highly likeness or eyes get sparkled

When to see some other expressions

When to end the conversation

Mon phayr na

Jab koi pasand na ho

Jab aankhain chundhya jain, ya buhat he pasan deedah kay nigah thahar na skaay

Kisi dosray kay expressions dhaikhnay kay liyay

Jab baat cheet khatam karne ho

Out

Outside weather is harsh.

Outside weather is harsh, windy, coarse and hot; Total investigative crucibility is to frame maximum energy usefulness.The procedures are required to manage natural processes such that minimized economic solutions can be attained as soon as possible.

Every possibility would be explored on each design step to enhance profit and yield. Rather working on a line, parallel group efforts would be essential to catch up the target; with mutual close timely work sharing.

Attitude

On the roads: burning tyres, destroying roads; breaking footpaths, streetlights, traffic signals, destroying government properties, destroying facilities and properties built from peoples' taxes, breaking peoples' properties on footpaths, in parking motorcycles cars shops 'chhabrie' 'rrayrrhees' destroying stocks; Who these are protesting presenting security threat to own society, Distructing country of Quaid and Allama.

Insanoon ko kaon tabah kar raha hay, baycar kar raha hay, moaashray kay leyay bojhh bana raha hay. Jo sirf abade main gintee hain.

Pillow sometime Pocket

There are very well spread within people of a group that this and that elders have miracles in their hands. From the chain of their 'Khadamain' personal close daily assistants, it is bubbled around to the ears of new arrivals in the group. This elder went to take a bath and he put a prepared empty pocket dress outside the bathroom; after sometime this elder came out and put the dress on, then there came a person asking some money and this elder put his hand in the empty pocket and when the hand appeared out of the pocket, there was few currency notes for that person. That elder went to deliver a 'biyaan' and after the 'biyaan' sat there rounded by the listeners; and he put a pillow on empty ground for that elder, then there came a person asking for some money and that elder put his hand under the pillow and pulled a few currency notes for that person.

Start

Queen of Yemen, picked her clothes up on the legs while in a meeting with Prophet of God King Sulaiman bin Daoud.

Wife of Prophet of God Ibrahim and mother of Prophet of God Ismail ran in between twin peaks of mountain marva and mountain safa.

Usually, people saw the Prophet of God in a 'tahbund' wearing from half body to under knees. In amamah, covering head, in kurta and in chaddar, toeing chaddar on the floor of his mosque to attend a guest in hurry. The Prophet liked 'shalwar' shape while seeing first Khalifa wearing it that is wearing cloth on two legs separately and suggested women to wear shalwar (8) under tahbund (o).

While standing or walking, first Khalifa was careful to hold his clothes up again and again, being careful about saving clothes.

Second Khalifa warned young people on to walk like or look like older people.

While on tour to Makkah on incidence of Hudaibiya as Prophet of God's representative, the third Khalifa wore clothes upto under knees. In era of third Khalifa, an old companion of Prophet of God hit hard on the leg wounding it as a matter of some warning.

Fourth Khalifa warned young people on to walk like women.

A companion of the Prophet of God of the same valley wore clothes on legs down due to thin legs.

Start

Queen of Yemen, picked her clothes up on the legs while in a meeting with Prophet of God King Sulaiman bin Daoud -Salaam on both-.

Wife of Prophet of God Ibrahim and mother of Prophet of God Ismail - Salaam on them - ran in between twin peaks of mountain marva and mountain safa.

Usually, people saw the Prophet of God Muhammad RasoolaAllah - peace be upon him and salaam - in a 'tahbund' wearing from half body to under knees. In amamah, covering head, in kurta and in chaddar etc. The Prophet liked 'shalwar' shape while saw the first Khalifa wearing it that is wearing cloth on two legs separately and suggested women to wear shalwar (oo) under tahbund (O).

Fourth Khalifa warned young people to walk like the women.

Saaq

Queen of Yemen, picked her clothes up on the legs while in a meeting with Prophet of God King Sulaiman bin Daoud.

Usually, people saw the Prophet of God in a 'tahbund' wearing from half body to under knees. In amamah, covering head, in kurta and in chaddar. Prophet liked 'shalwar' shape while saw first Khalifa wearing it that is wearing cloth on two legs separately and suggested women to wear shalwar under tahbund.

While on tour to Makkah on the occasion of Hudaibiya as Prophet of God's representative, the third Khalifa wore clothes upto under knees. In the era of the third Khalifa, an old companion of the Prophet of God hit hard on the leg wounding it as a matter of some warning.

Shine

On the day of Qiyamat, God will reveal his 'Saaq' and the people will be unconscious by the scene.

Prophet of God Muhammad Rasoolallah when will become awaken will see Prophet of God Moses embracing the pillar of throne of God. It is told that whether he would become awaken earlier or never would become unconscious due to his earlier unconsiousness on the mountain of toor.

It is written somewhere that the 'hoor', the woman of Paradise would be so beautiful that her legs would be clean clear neat transparent so the plasma in bones would be seen.

Shape and Face

God does not see shape or face or any one but the acts and deeds - neat in hearts - that is God sees the heart. Every heart is within the fingers of God and whenever God wants might change it.

Some prayers are thrown back on the face of the prayerer. Some reciters do not understand the Quran and the quran does not go to their heart while they are reciting it a lot. Reciters of the Quran are asked to 'witr' prayer five units or seven units at night.

God will come in a face so that people will ask who it is? and Angels of God will tell them that it is their God (your Rub) and then the people will recognize and say ' Huq' that is truth upon telling the Angels about peoples' God.

Afraidness

The afraidness of God will be a source of forgiveness. A person before the time of Islamic Prophet of God upon asking told that while before God that he advised to burn his dead body due to afraidness of God whereupon God forgives him. It happened on his death or will be on the day of Qiyamat. It is possible for God. Grave is a door to 'aalam of akhirat'. In a house - 'house' is a word - there is corridor, yard, garden, bedroom, bathroom, toilet, roof, stairs, balcony, kitchen, storeroom, dining room, meeting room, drawing room, and many other rooms to mention. So as so the grave is - the 'grave' is a word - for ' illiyeen, safaleen, Paradise palaces, Paradise gardens, hell apartments, hell wells etc.

Communication

God communicated with humans through his Prophets of God - Salaam on them-. God might talk directly or indirectly to selected Prophets of God. Via a tree God talked to the Prophet. The Angel of God of Talk between the Prophet and God is appointed and always reached to the Prophets with messages of God. There are many messages of God and books of God. The last and final Book of God Al-Quran was revealed to the Prophet of God Muhammad Rasoolallah in a time period of about twenty three years as a word, verse, many verses, a part or repeating a verse again or changing a verse with other verse. Overall, the Prophet of God does not word anything without the 'wahi' talk of God. The Prophet's dreams are also a truth way of communication. God talked to the Prophet of God face to face saying come closer, distance was two bows or less than that. Fourth Khalifa while on gathering prayer announced on the 'minbar' climbed with a bow in hand. On the sky, for an observer on the earth the two stars look closer having a distance of less than a finger width, however, in actual there is a considerable huge distance in between. The Quran, the Book of God, revealed on the Prophet of God to act upon it while living on the earth atmosphere. This is for or from whom who was seeing the meeting.

Dream

In a dream Prophet of God Muhammad RasoolaAllah - peace be upon him and salaam - saw a white palace and while entering told by the Angels of God that it was of Second Khalifa. The Prophet of God Muhammad RasoolaAllah - peace be upon him and salaam - also saw Paradise blessings and hell punishments. While praying once, the Prophet of God Muhammad RasoolaAllah - peace be upon him and salaam - saw gardens of Paradise and stepped ahead to grip. Dreams are seen, if there are thousands of peoples are sleeping, by one or two, by a few numbers, but it is not said that hundreds of people don't see dreams so there is no reality about dreams. Dream of the Prophet of God is a way of communication of God to the Prophet. God communicates to honeybees to train honey bees to do the work of honey. People of mind rank on any way also communicate mindly to each other. Dreams come true and it is a matter of time. The learning of dream answering is a science and trained people have the knowledge of answering and understanding the dreams. Philosophy holds here and it is not a reason to act upon dreams as it is. Dreams may be from good or bad. Dreams may tell future or present situation solutions with the mind power. However, a poise person dreams of the truth. Guidelines may be provided by nature to them by true dreams. A mistake in dream interpretation might happen. And God always knows the best.

Private

The Prophet of God Muhammad RasoolaAllah - peace be upon him and salaam - and his family have many things that are being narrated but are private. There are many incidents reported. Arm, leg unclothing and such rare happened incidences been reported in history are how to be acted upon is a matter of wisdom. The time of incidence, place of incidence - whether it is inside house or outside house, it is with close companions, relatives or with public companions -.

There is a lot of difference between man and woman : physically that also influences upon their daily work routines, responsibilities and living. That is also cared for while asking them for prayer. Everything whether the woman has to do as of the Prophet or the woman is told something else. There is a difference of orders of works as the order of man and woman. Then the Prophet sometimes sat bare head with hair combed from in between. During a prayer, the Prophet up the hands so high that the whiteness of armpits was seen.

Praying physical prayer is good for women in the house yard and very good in the house and better in the room and so.

Private

There is a difference of orders of works as the order of man and woman. Then the Prophet sometimes sat bare head with hair combed from in between. During a prayer, the Prophet up the hands so high that the whiteness of armpits was seen. There is a lot of difference between man and woman : physically that also influences upon their daily work routines, responsibilities and living. That is also cared for while asking them for prayer. Everything whether the woman has to do as of the Prophet or the woman is told something else. There is a difference of orders of works as the order of man and woman. Then the Prophet sometimes sat bare head with hair combed from in between. During a prayer, the Prophet up the hands so high that the whiteness of armpits was seen. Once inside the house the Prophet of God was sitting bare knee with his two father in law. There are so many incidents which differentiate between men's and women's and it is a wisdom.

Woman

Women are not allowed to talk behind the leader of prayer during prayer to indicate to the leader but with clapping - they are allowed to pray in mosques with restrictions and a protected environment.

Public

In public, there are many things which are automatically become restricted while those are allowed in private and vice versa. Showing the ankles, legs are allowed for men.

Boundaries, time restrictions, prerequisites, place - home, bazaar, public place, work areas etc. - health, time, gender, alone or growd, travelling or staying, area with laws and rules, Islamic areas, non-islamic areas, houses, shops, work areas, each and every aspect demands different styles of living keeping in view the right ways to please God.

Decline of Spain and Hindustan and other areas where the rule was Islamic land but now is not is a matter of ignorance and acting upon own wishes rather than obeying orders of God and Prophet of God Muhammad RasoolaAllah - peace be upon him and salaam - in a right way as done by companions of the Prophet.

Reasons

Forgetness, sickness, meals - chicken, goat, meat eating with using knife- use of right and left both hands to pick up pot to drink or eat, journey, business, day and night shift works, cloths, face and shape, beauty, gorgeousness and youngness and many other things are the variables for sorting the orders of doing and obeying.

Headache

Wound - headache - in head and/ or as water is not available for cleanliness, bath or wazu, then it is allowed to do 'tayammum' with neat and clean dust.

On the day of Qiyamat, God when would asked to sit on the prescribed place for the Prophet of God, the Prophet will put a hand on it while standing very anxious about his Ummah.

Subhanallah

Subhanallah for men behind the leader of prayer to point to the leader of any mistake whereas women are allowed to clap to caution the leader of prayer. It is allowed an equivalent of Subhanallah for the late comers to join the prayer with the leader of prayer saying equivalent Allah o akbar. Prophet of God Isa (salaam on him) on the day of Qiyamat will say Subhanallah to the One asking about his life on the earth. And God knows the best.

Likeness

The Prophet of God and his family have many things that are being narrated but are private. There are many incidents reported. Same is the case that there are many things which are left on their own choice. The demand of Islamic laws is to establish peace at every corner of the ruling areas and 'death of a hungry dog on the shore of a river is the responsibility of the government' that is not only peace but the food too. Islam is a religion of peace and to offer peacefulness to every life on the earth sothat each and every one can worship God peacefully having rights to live on te earth without under some cruelness of other peoples/ rulers.

The Prophet of God Muhammad RasoolaAllah - peace be upon him and salaam - once sitting with his companions told his three likelinesses. Whereas the first Khalifa told his own three choices. Second Khalifa told his own three choices. Third Khalifa told his own three choices. Fourth Khalifa told his own three choices. The Angel of God then arrived there and told his own three choices and also told the three choices of likeliness of God. This is for Ummah to have an open choice of living while acting and obeying God and Prophet of God.

Reasons

Forgetness, sickness, meals - chicken, goat, meat eating with using knife- use of right and left both hands to pick up pot to drink or eat, journey, business, day and night shift works, cloths, face and shape, beauty, gorgeousness and youngness and many other things are the variables for sorting the orders of doing and obeying.

Public

The Prophet of God Muhammad RasoolaAllah - peace be upon him and salaam - and his family have many things that are being narrated but are private. There are many incidents reported. Same is the case that there are many things which are left on their own choice. The demand of Islamic laws is to establish peace at every corner of the ruling areas and 'death of a hungry dog on the shore of a river is the responsibility of the government' that is not only peace but the food too. Islam is a religion of peace and to offer peacefulness to every life on the earth sothat each and every one can worship God peacefully having rights to live on te earth without under some cruelness of other peoples/ rulers.

The Prophet of God Muhammad RasoolaAllah - peace be upon him and salaam - once sitting with his companions told his three likelinesses. Whereas the first Khalifa told his own three choices. Second Khalifa told his own three choices. Third Khalifa told his own three choices. Fourth Khalifa told his own three choices. The Angel of God then arrived there and told his own three choices and also told the three choices of likeliness of God. This is for Ummah to have an open choice of living while acting and obeying God and Prophet of God.

Public

In public, there are many things which are automatically become restricted while those are allowed in private and vice versa. Showing the ankles, legs are allowed for men.

Boundaries, time restrictions, prerequisites, place - home, bazaar, public place, work areas etc. - health, time, gender, alone or growd, travelling or staying, area with laws and rules, Islamic areas, non-islamic areas, houses, shops, work areas, each and every aspect demands different styles of living keeping in view the right ways to please God.

Decline of Spain and Hindustan and other areas where the rule was Islamic land but now is not is a matter of ignorance and acting upon own wishes rather than obeying orders of God and Prophet of God Muhammad RasoolaAllah - peace be upon him and salaam - in a right way as done by companions of the Prophet Muhammad RasoolaAllah - peace be upon him and salaam -.

Some of true Islamic rulers did not stay longer because of their strictness on things which were not either compulsory for the society or local scenario demanded to be ignored for sometime was not cared for, or their sitting level with opposition from ownself based on piousness was very high.

Amble

Little sorrow or grief to humans nullify the bad deeds accordingly. Good deeds and bad deeds are accumulating on the humans (and the jinns) for their examination on the day of judgment at the time of Qiyamat. Some people get benefitted from their good in the world and at the same time sorrow and grief against their bad deeds. Companions of the Prophet of God alert themselves on worldly successes whether they were getting the reward in the world only. For pious persons sorrow and grief enhances the ranks in the court of God and in Paradise. There will be people in between good and bad having equal good and equal bad nullifying each other with zero to go to paradise or hell on the day of judgment. In the world whether their good are benefited or their bads are punished making an equal good and bad. They will wait until permission of God to go to Paradise.

The earth life had been started by the pair of the human Adam - Salaam on him - and his spouse Hawwa - salaam on her - who was created by God Al-mighty from the left side of Adam - Salaam on him - at skies. The process of creation of the human Adam was told by God himself via his Angel to the Prophets of God for the human generation. God was in a hurry to make Khalifa on the earth. God talked with the audiences present before God in the Court of God and those were Angels of God, the jinns, the human and other creations God knows.

God created the human body, container of soul, from the collected chemical materials from earth by Angel of God. Everybody is injected by a soul by God. God created the second human body from the first human. God designed the human structure in its best shape. In the sky-system, God has the throne on the water, the Chair of God covers the earth and skies.

Angels of God have many assigned duties by God. An Angel is the messenger of God between God and the Prophets of God and also delivers punishments of

God on the non-believers. An Angel is on delivering rain and food to all creations. People were used to hate one and like the other. God does not like to differ among Angels in case of believing them. God does not like to differ among the Prophets of God in case of believing them. Prophet Adam who was created Khalifa for on the earth with his pair Hawwa both were transferred to paradise-stay in the sky-system. They both got proximate with the 'shajarah' there and then transported to the non-skysystem on the earth. The line between sky-system and non-skysystem is a port whereas any line drawn in a system upon which some command restrains is fer and in the paradise-system it is shajarah - from the court of God to paradise, it is the gate of paradise in the sky-system.

Prayer is the key to the gate of paradise. The jinn stated that the jinn would capture and deceive all the humans with exceptions. Prophets of God told humans (and jinns) the Path of God. God is ever powerful. The challenge of the jinn was depicting that the knowledge of the humans to go to the earth was predicted by the jinn whereupon the jinn asked for permission from God to do work of deceiving people from right path of coming back to paradise till a fix time known that is time of Qiyamat.

It is education, knowledge and reading and writing and the knowledge of the unknown is Science. As the knowledge expands it branches into many departments based on similar things and properties. Life of a human is very short to cover up now so expanded knowledge in many disciples since the human creation by hundreds and thousands intellectuals in respective fields.

Both Adam - salaam on him - and Hawwa - salaam on her - were asked to stay at Paradise and enjoy their eating, drinking there, and here God commanded them not to cross the 'shajarah'. The orders of God contained the allowed acts and the refrained acts. Before that the creations at sky-system were commanded by God to prostrate towards the human Adam that is in sight-seeing prostrate to Adam. It was a command of God. The jinn (Satan) denied and God declared the jinn 'Shaitan Rajeem'. The jinn became against humans, prouding on his creation from fire whereas humans were created from earth materials, and asked for permission to deceive humans (and their children) rather than sorry before God repentantly.

God commanded both Adam - salaam on him - and Hawwa - salaam on her - enjoy at Paradise and not to cross the 'shajarah'. The orders of God contained the allowed acts and the refrained acts. These commands are God knows for the time to time changed, altered and renewed by God. Angel of God revealed these commands via messages of God, Books of God to the Prophets of God till the last and final Book of God Al-Quran.

Evolution of Commands

Prophet of God Adam - salaam on him - and Hawwa - salaam on her - were asked to stay at Paradise and enjoy their eating, drinking there, and here God commanded them not to cross the 'shajarah'. The orders of God contained the allowed acts and the refrained acts. Women had gotten a grief of monthly menstruation which not only clears the body internally but also clears the bad on her. Every sorrow or grief, how little it is, nullifies bad otherwise adds good.

Before that the creations at sky-system were commanded by God to prostrate towards the human Adam that is in sight-seeing prostrate to Adam. It was a command of God. The jinn (Satan) denied and God declared the jinn 'Shaitan Rajeem'. The jinn became against humans, prouding on his creation from fire whereas humans were created from earth materials, and asked for permission to deceive humans (and their children) rather than sorry before God repentantly.

God commanded both Adam - salaam on him - and Hawwa - salaam on her - enjoy at Paradise and not to cross the 'shajarah'. The orders of God contained the allowed acts and the refrained acts.

Time by time as the humans lived on the earth, God changed, altered and newed commands for them. It was a reason to test them. Whether in hardships or in ease. Commands of God are related to human acts. God time by time as God wishes upon humans sends his blessing. God revealed good and bad both on the humans to ask them to be clean clear neat themselves. God creates death from life and life from death. God is all alone to do so. God is one. God is independent. Nothing is from God and God is from nothing. Neither God is born from anyone nor anyone born from God. Neither God is separated born from anyone nor anyone separated born from God. God has no partner, no spouse, no wife, no equivalent, no mate; God is one. God never has any 'kuffoo'. God is from ever. Adam - salaam on him - was created.

Hawwa is from Adam. Every human is from humans. Isa (Jesus) is from Maryam (Mary) born as Hawwa was created from Adam. Adam was from earth material and there was nothing human before Adam. There was Maryam (Mary) before Isa (Jesus). By the command of God, Adam, Father of Humans, came in life.

At the start of life on earth marriages were between male and female born at different times for the very first pair of Adam and Hawwa - salaam on both of them -. And it was a command of God. By the passage of time, the order regarding marriages were changed, altered and renewed by God through Prophets of God. It is now Religion Islam of the Prophet of God Muhammad Rasoollah Nabiallah Habiballah Abdullah - Peace be upon him and salaam - who is the last and final Prophet of God for humans (and the jinns) till the Qiyamat including days of Qiyamat, however door of repentance will be closed at the start of Qiyamat before Qiyamat, which is God selected for the humans (and the jinns). Accountability of humans has become closer and they are Ghafal and in Aaraaz having no feeling to remember God. The Religion (evolution of commands) for humans (and the jinns) of the humans (and the jinns) has been completed up, the blessing of God upon the humans are total and God is pleased with you on Religion Islam on the day at the time of Huj of last and final Prophet of Islam Muhammad Rasoollah Nabiallah Habiballah Abdullah - peace be upon him and salaam -.

With the passage of time on earth, the Prophets of God called the humans towards God, reminding them their origin and the Paradise. From Adam - salaam on him - the Prophet of God taught his childrens the commands of God to live on the earth. Both Adam and Hawwa were sent to earth after proximation of shajarah and there they both were taught the prayer of repentance to pray before God. by God via Angel of God. Then Angel of God revealed the message of God to them and the prayer of repentance was the first message for them on the earth. First command was on the sky of living in Paradise. Second to come on the earth.

The command of God to live and enjoy in Paradise for them. There will be a time after day of judgment when on Qayamat good will be in Paradise and bad will be in hell. And there will be a loud call from the side of God asking Adam (Father of Humans) to turn off the fire from his children that is to pray the prayer, whether it will be the first prayer taught to Adam or the prayer would be taught to Adam thereupon; as the Prophet of God Muhammad Rasoolallah Nabiallah Habiballah Abdullah - peace be upon him and salaam - will be taught on the day of Qiyamat.

Qiyamat be comprised of the day of the start of qiyamat, Day of God, day of judgment.

Qabar: On death time it start till Hashar. Life of Barzakh.

Hashar: Every dead life will be relifed as of being body on the earth.

Meezaan: day of accountability and judgment

Chapter Messages

اعوذ بالله

انی جاعل فی الأرض خلیفة

الله

From the day first of the human life world, God has changed, altered, renewed many of his commands, orders and messages. From the first messages to the last and final Book of God Al-Quran, each is on the same level in case of Iman on Books of God. The essence of all previous messages and Books of God is in the last and final Book of God Al-Quran. It contains everything needful for humanity till day of Qiyamat including talks of God to with his creations, dialogues at skies, acts of past nations, punishment to past nations, worldly things, sky things, universe, praise of God, mention of Prophets, prayers, commands, materials, animals, Paradise, Hell, Qiyamat, day of judgment, talks of peoples to mention.

All Prophets of God from Prophet of Adam - salaam on him -, to the last and final Prophet of God Muhammad Rasoolallah - peace be upon him and salaam - are equal on the same level in case of Iman on Prophets of God. Prophets of God were the same on saying and believing God is one, having differences in time periods, time span, age, nations, people following obeying count - disobeying count, rule, journeys, family, wars, and many other things. In the skies they are in different skies. The top most Prophet of God is the last and final Prophet of God Muhammad Rasoolallah - peace be upon him and salaam - who by the power of God travelled from Makkah Masjid Haram to Masjid Aqsa and there the last and final Prophet of God Muhammad Rasoolallah - peace be upon him and salaam - leaded prayer to all the Prophets of God who were also there by the power of God.

God commanded both Adam - salaam on him - and Hawwa - salaam on her - enjoy at Paradise and not to cross the 'shajarah'. The orders of God contained the allowed acts and the refrained acts.

The message of God to both was direct to them. Then they came on the earth. They were created to live a life on the earth. Satan succeeded to lead them to earth by deceiving them of proximity with 'Shajarah'. After their struggle at earth they got a message of God via Angel of God to repent and pray. They accepted their act as against command of God and repented and forgiveness and blessing of God covered them.

Prophets of God were chosen from among humans at the very early promise declaration of all souls of humans in the sky. God created Adam. All souls from the first to last human before Qiyamat who has to become in life on the earth were created from Adam by the command of God. God asked them about their Rub. And each and every soul said yes, why not! confirming God being the Rub. God is Rub ul Aalameen. Parents are Rub while their children are little. At forty years of age they are asked to pray for 'rehem' for their parents, O Rub mercy on both as they rub us when we were younger less than forty years of age. The love and affection for children are provided in the hearts of parents by God.

Government at any land area is a rub to the people and creation living there. Second Khalifa took responsibility by saying that he would be responsible for a hungry dog within his area land of rulership. God has made different levels of humans so that the earthy life can be governed accordingly. Whether there are two Gods in the skies, there would be huge wars and quarrels among them. There is only one God all alone. God is managing the whole universe with his commands and orders via Angels of God. The messages of God were reached to Humans via Angels of God to the Prophets of God. All are in protection of God. God what God likes retained from his messages and what God likes told to the Prophets of God. The ordered messages of God contained the allowed acts and the refrained acts. These commanded messages are, God knows, for the time to time changed, altered and renewed by God. Angel of God revealed these commands via messages of God, Books of God to the Prophets of God till the last and final Book of God Al-Quran to the last and final Prophet of God Muhammad Rasoollah Nabiallah Habiballah Abdullah Rehmat ul liaalameen - peace be upon him and salaam -.

Pious Ways

God was in a hurry to make Khalifa on the earth. God created earth and slides within six days, God knows. Prophet Adam who was created Khalifa for on the earth with his pair Hawwa both were while in paradise-stay before coming to earth got proximate with the 'shajarah'. There they were finding something to hide themselves. Then they left the sky-system to come on the earth. God transported them from sky to earth and delivered clothes following them. Then humans found earth plants to make thread and clothes. Clothes signify the profession and the best clothes are piousness. God set earth life for humans creating life and death to testify humans (and jinns) for the best actions. God Prophets of God reminded humans (and jinns) that God had created them and everything. They are here on the earth to recognize God to worship God.

The Prophet of God was given followers by God and got companions from the people who were first weak then grew up stronger. The friends of God (owl-le-ya) in here and in hereafter - don't become sorrowful and are not afraid. The Prophet of God is first for believers upon their own selves and friends are special cases. Green fields are liked by farmers. Green field then turns blackish.

The Prophet of God said about the immigrants as his outer clothes whereas about supporters as his inner clothes. Husband and wife are both clothes of each other. Best clothes are pious. The Prophet of God wore a cloth sent as a gift for him from an outer land and then did not wear it again. The Prophet of God did not like to eat an animal from outer land. The Prophet of God abandoned the coming of people to the Mosque of the Prophet of God who ate some vegetable-fruity-crop of opposition. Different levels of people are so that worldly works can be in progress. Good is working, evil is working while living together in this world. Day of God will sort out good to paradise and evil to hell.

Pumpkin Ways

God was in a hurry to make Khalifa on the earth. God set earth life for humans creating life and death to testify humans (and jinns) for the best actions. God reminded humans and jinns about God and Day of God via Angels of God, Books of God and Prophets of God. First Prophet of God was the first human and that was Prophet Adam who forgot not to proximate 'shajarah' with his pair and both had arrived on the earth after that from the sky-system. God forgets who forgets God. God knows. The forgetful style of God is known to God. God never takes sleep or catnap. God waits to forgive the mistaken-er on their mistakes. Life on earth is a journey to the Day of God. Lives are prisoners here being captive in snare. All Prophets of God were men. Prophet Adam and his pair did hardships remembering God and finally got God forgiveness. Prophet of God Yunus went away from his nation forgetting them and was being trapped in storms in rivers and then in fish belly to on the shore covered under shadow of a pumpkin vine (bail) so forgiveness of God attended him and he learnt spells from God to remember God after his forgetfulness. Forty years in deserts, the tribes of the Prophet of God Moses were forgotten. Prophet of God Isa will say Subhanallah upon asking of God whether he said to people to make him and his mother gods or so and this is a judgment of the people and the talk is on the Day of God. Reminding the leader in prayer when he forgets is saying Subhanallah for by men and is hand on hand hitting for by women. Also, note, the caller of prayer and the leader of prayer must always be a man.

Process Ways

God was in a hurry to make Khalifa on the earth. At the time of the announcement of Khalifa there were angels, jinns, and many other creations, God knows. God commanded to prostrate bowing to the floor towards the human. Angels obeyed whereas the jinn denied causing that the jinn was created from fire whereas the human was created from mud. God knows the hearts. God created life and death to testify who does better actions. Prophet Adam who was created Khalifa for on the earth with his pair Hawwa both were while in paradise-stay before coming to earth got proximate with the 'shajarah' and then left the sky-system to come on the earth. That was a journey from sky to earth, God knows. They both came on the earth and human life started. Life and death was created by God to test who does the best actions. Prophets of God continued to remind and link the public to about God, Angels of God, Books of God, Prophets of God and Day of God till the last and final Prophet of God.

Prophet of God and his companion who was accompanying the Prophet of God as per likening of the Prophet of God with permission of God. Both when they arrived at Madina coming from the city of Makkah, people at Madina welcomed them. These people were the first supporters and so they have seen the Prophet of God and his first companion. At Madina there were tribes and groups of public who had been in the process of deciding their ruler. They were ready to crown a head of their state at the nip time of the arrival of the Prophet of God. Public and leaders of different tribes rushed towards the Prophet of God and without any hesitation majority of the people and tribes accepted the Prophet of God their everything from heart and soul. At the time of departure of the

Prophet of God from worldly life to God - life and death is created by God to testify humans (and the jinns) who do the best action-reaction. When a first among firsts was said to ask for ruler-ship participation, the reply was that if denied by the Prophet of God upon asking then the people would never offer them in future. People remembered the very arrival of the Prophet of God and the decade life of the Prophet of God with them and whoever they liked they crowned, God knows, first caliph of Islam, Khalifa of Prophet of God - who was firm and hard enough on the spot where the Prophet of God left the ummah at his departure from world to God -.

Smooth Ways

God was in a hurry to make Khalifa on the earth. God set earth life for humans creating life and death to testify humans (and jinns) for better actions. God reminded humans and jinns about God and Day of God via Angels of God, Books of God and Prophets of God. First Prophet of God was the first human and that was Prophet Adam. The list ends at the last and final Prophet of God Muhammad Rasoolallah (Peace be upon him and Salaam on all the Prophets of God). During the time of each Prophet of God who ever followed the Prophet of God was good and successful, and who ever opposed the Prophet of God was evil and failed.

God started life and death on the earth to testify humans (and jinns) who are on better actions. God sees not faces or outer skins but the heart and the actions. Best is who is more reluctant to evil. God has created tribes and classes among humans (and jinns) for intro and it is used for population balance, reorganizations and reconciliations. God has created languages so the humans can explore earth and to make their living on earth. Humans (and jinns) are created for worship of God and it is after recognition of God through creations of God around humans (and jinns) on the earth and in the skies. Prophets of

God guided them to the right path whenever it was required by God that humans (and the jinns) needed, God knows. God put humans on different levels so the worldly works can be done smoothly otherwise society may be stir (mooth) with injustice and wrongs ending to the law-less environment. Standing one has an upper level on sitting ones that is to go for to do any good for God sitting one has to stand. It would be hard doing any work when each has a similar level. Doing one-another work requires that there are life levels. Good will get the promise of God that is paradise and evil will also get the promise of God that is hell.

God was in a hurry to make Khalifa on the earth. God set earth life for humans creating life and death to testify humans (and jinns) for better actions. God reminded humans and jinns about God and Day of God via Angels of God, Books of God and Prophets of God. First Prophet of God was the first human and that was Prophet Adam. The list ends at the last and final Prophet of God (Peace be upon him and on all the Prophets of God). During the time of each Prophet of God who ever followed the Prophet of God was good and successful, and who ever opposed the Prophet of God was evil and failed.

God started life and death on the earth to testify humans (and jinns) who are on better actions. God sees not faces or outer skins but the heart and the actions. Best is who is more reluctant to evil. God has created tribes and classes of humans (and jinns) just for intro and recognizations. God has created languages so the humans can explore earth and to make their living on earth. Humans (and jinns) are created for worship of God and it is after recognition of God through creations of God around humans (and jinns) on the earth and in the skies. God put humans on different levels so the worldly works can be done smoothly. It would be hard doing any work when each has a similar level. Doing one-another work requires that there are life levels. Good will get the promise of God that is paradise and evil will also get the promise of God that is hell.

The last and final Prophet of God is Prophet of God and his relations are not counted in any way to success else one has good and as permitted by God. God said to the Prophet of God that love of the Prophet of God to anyone does not guarantee of righteousness but righteousness is with God. God guides to ways of God for who struggles for God in God for God. Prophet of God Muhammad - peace be upon him, on his family, on his companions, on his wives and on his ameers and salaams - is not father of any man among the people at the time of Prophet so as a straight rule of ruler-ship that is son is the successor don't hold. The Prophet of God left peoples to decide for his successor and they immediately did as God liked. First caliph appointed his successor and it was also with mutual consultations. Second caliph after mutual consultation appointed a committee of six members (co-sush) to select one from among of them. Fourth caliph was also the member of the second caliph's committee and people asked to him to take the responsibility then directly.

Chapter Computation

اعوذ بالله

انی جاعل فی الأرض خلیفة

الله

God was in a hurry to make Khalifa on the earth. God created the human and commanded the personalities present in the court sky-system of God to prostrate bowing to the floor before the first human. Angels of God accepted the command and prostrated whereas the jinn rejected the command and did not prostrate. There were then God stayed the humans in paradise and ordered the jinn to get out whereupon the jinn asked for permission of deceiving the humans till a fixed time limit, God knows. Where the jinn would deceive the humans that were on the earth and the jinn, firstly, also succeeded during their paradise-stay. Life started on the earth. The humans are exploring earth to recognize their creator whereas the jinn is showing them the seen things, creations, and surface realities being their God. God sent Prophets of God to remind God and Day of God. Angels of God delivered Books of God to Prophets of God for the guidelines on the earth. The final religion named Islam was liked by God for the humans. Prophet of God Ibrahim said that he was among the first Muslims. The message of God from the first Prophet of God to the last and final Prophet of God is the same that God is one and there will be a Day of God. Time by time, based on earth situations and human nature development, God altered, changed his other commands of routine life on the earth for the humans. Quran - the Book of God - having every essence of what the God had revealed to his Prophets is the Book of God till the Day of God. Day of God came closer. Very close is the day of judgment of the people. A week has seven days. Friday is close means today is either Thursday or

Wednesday or rarely Tuesday. Before Qiyamat, the doors of forgiveness will be closed symbolically. Qiyamat - Day of God, Day of Judgment and Time to sort good to Paradise and evil to Hell, if said that is coming close then knowing that the Day will be fifty thousand days and in regular there will be each day equivalent to thousand days. The announcement of closeness of the day is either equivalent to thousand, two thousands or three thousands years or and in case of basing fifty thousands the announcement of closeness of the day is either fifty thousand, hundred thousands or so, God knows.

A narration from the Prophet of God mentions that the present ummah time is equivalent to the last day part. So, in the case of a thousand years a day means it would be (157 - 167 - 177) upon double shadow and (314 - 324 - 334) upon single shadow and approximately less than around half of the day ~ 500 for some, but these figures have been passed. Relating it to the fifty, it will be (7850 - 8350 - 8850) years upon double shadow and upon single shadow it will be (15700 - 16200 - 16700) years exceeding less than to twenty five thousand, God knows.

For more verification a population balance may be tested upon. On the day of Qiyamat, out of 120 lines there will be eighty lines of the present ummah of the Prophet of God containing Muslims as well as non-muslims! In the case of Muslims only, the day is far away longer than in case of both, God knows.

Postface

The creations at the sky-system were commanded by God to prostrate towards the human Adam that is in sight-seeing prostrate to Adam. It was a command of God. The jinn (Satan) denied and God declared the jinn 'Shaitan Rajeem'. The jinn became against humans, prouding on his creation from fire whereas humans were created from earth materials, and asked for permission to deceive humans (and their children) rather than sorry before God repentantly.

God commanded both Adam - salaam on him - and Hawwa - salaam on her - enjoy at Paradise and not to cross the 'shajarah'. The orders of God contained the allowed acts and the refrained acts. The allowed acta are called 'halaal', and refrained acts are called 'haraam'. Allows sometimes become refrained and refrains sometimes become allowed. It is wisdom of orders and acts ported and explained by the Prophet of God along with teaching of cleanliness of not only body and environment but neatness of heart.

In public, there are many things which are automatically become restricted while those are allowed in private and vice versa. Showing the ankles, legs are allowed for men.

Boundaries, time restrictions, prerequisites, place - home, bazaar, public place, work areas etc. - health, time, gender, alone or growd, travelling or staying, area with laws and rules, Islamic areas, non-islamic areas, houses, shops, work areas, each and every aspect demands different styles of living keeping in view the right ways to please God.

It is education, knowledge and reading and writing and the knowledge of the unknown is Science. As the knowledge expands it branches into many departments based on similar things and properties and spread of knowledge in many disciples since the human creation by hundreds and thousands intellectuals in each respective field is huge.

Totalday that - God - is in a ShaN.

Whom the humans (and the jinns) are trying to deceive! God Al-mighty has out-appointed away the jinn (Satan) to test the humans (and the jinns) for - to come back to Paradise or otherwise for hell. Humans need lands and sources, and upon these they quarrel. Who is righteous and who is obeying Satan is also known to God. Satan will stand aside away on the third day of Qiyamat whether some believe it or not. Habit of snakes is not to be blamed. Hell is to who denies. The erudite escapade escalations are rambled so that such, what it is! can be drawn. It is a book. It is easy and simple but is disabled on some minds for years. The link is on but the data entry is not.

Prophet of God Muhammad - Peace be upon him and salaam - encouraged local products, local industry and discouraged import.

Fourth Khalifa put emphasis on shoora and cared for public opinion.

Wisdom is the basis to attend the commands of God.

References

-Book of God, Al-Quran

-Book of Imam Bukhari, Sahih Al-Bukhari

-Explanations of the Book of God Al-Quran, Explanations on the Books of God, translations and historic books regarding Islamic history, Sahah Sitah, Moata

-Books on Imam Abu-Hanifa

-Arabic Urdu Dictionary

-Oxford, Advanced Learner's Dictionary, Fourth Edition

-The following short articles in the book are the essence of my long time readings and thoughts having seeds in my mind and my diaries. Zahid

-Mixing in pipelines with side-tees, Zahid Hafeez Khokhar, HD Zughbi, R Sharma, The 6th Saudi Engineering Conference, KFUPM, Dhahran, December 2002

-Numerical and experimental investigations of mixing at side and opposed-tees, HD Zughbi, Zahid Hafeez Khokhar, RN Sharma, AIChE annual conf., Indianapolis, USA

-Investigations of mixing in pipelines with side-, opposed-, and multiple-tees, Zahid Hafeez Khokhar, King Fahd University of Petroleum and Minerals, 2002

-Investigations of mixing in pipelines with side-, opposed-, and multiple-tees, Zahid Hafeez Khokhar, King Fahd University of Petroleum and Minerals,

-Investigations of jet injection arrangements into main flow, Zahid Hafeez Khokhar, H Zughbi, S Sadiquee, Sharma RN, International conference on chemical and bioprocess engineering, ICCBPE, Aug 2003

-Effects of jet arrangement on mixing in pipelines with side-tees, Zahid Hafeez Khokhar, HD Zughbi, S Siddiqui, International conference on chemical and bioprocess engineering, Universiti ...

-Decarburization of Fe/C melts in a crucible: Effects of gas flow rate and composition, HD Zughbi, Scandinavian journal of metallurgy 32 (4), 194-202, 2003

Bibliography

Zahid Hafeez Imam Khokhar, (Zahid Hafeez PEC-3399-Chemical Engineering-1998, P. Eng.), Pakistan. Chemical Engg. B.Sc. (Session 1992-96). Chemical Engg. M.Sc.(2002) / ... Professional Activities: Lecturer, Assistant Engineer Officer (Design Process), Research Assistant, Co-investigator, Business Development Manager, Visiting Lecturer, Lecturer, Publications

Zahid Hafeez Imam Khokhar, (Zahid Hafeez PEC-3399-Chemical Engineering-1998, P. Eng.), Pakistan. Chemical Engg. B.Sc. (Session 1992-96). Chemical Engg. M.Sc.(2002). Professional Activities: Lecturer, Assistant Engineer Officer (Design Process), Research Assistant, Co-investigator, Business Development Manager, Visiting Lecturer, Lecturer, Visiting Professor, Research & Development, Publications.

MIX
Papier aus verantwortungsvollen Quellen
Paper from responsible sources
FSC® C105338

Printed by Books on Demand GmbH, Norderstedt / Germany